THE FOOD FOREST
NOT A FANTASY
As told by Professor Tola

The Food Forest, Not a Fantasy - As Told by Professor Tola
Written by Lisa Cacao and Illustrated by Phyllis Thiessen

Published by Tola Learning Press, Warner Robins, Georgia, USA
Paperback Edition
ISBN: 979-8-9963075-3-1
Library of Congress Control Number: 2026913659

This illustrated children's book is a work of creative educational storytelling intended to inspire environmental stewardship, ecological awareness, and appreciation for creation through imaginative narrative and artistic expression.
Printed in the United States of America
First Paperback Edition

For information regarding permissions, educational use, or future titles from Tola Learning Press, please contact the publisher at tolalearningpress@outlook.com.

For יהוה (Yahuah),
for inspiring and teaching me through this process.
May You be glorified by the work of my hands and by the creativity You have placed within me.

With great appreciation and love for my husband, Javier, for your patience with me as I worked on this book.

For my boys, who continue to inspire me to return to being creative.

And with heartfelt gratitude to Phyllis,
for your dedication to preserving the integrity of this story, and for the prayers, care, and research you poured into understanding the life of a worm so beautifully.
Thank you for helping bring this vision to life.

Endorsements

In a time when the natural order is being forgotten, The Food Forest - Not A Fantasy, revives an ancient truth: that the Earth, when stewarded with wisdom, yields abundance. This is not merely a creative book about sustainable agriculture; it is an inspired call to return to divine design, to systems of living that honor both creation and Creator. Through the voice of Professor Tola, this work delivers profound insight wrapped in whimsical yet accessible narrative.
— Dr. Stephen Pidgeon, Author, Educator and Translator, The Eth Cepher Collection

The Food Forest - Not A Fantasy is a beautiful, hope-filled book that invites children to see the land as a source of life, abundance, and purpose. With warmth and imagination, it inspires young readers to care for creation and see themselves as part of its restoration. This is the kind of book that stays with you, and makes you want to dig in and help grow a better world.
— Chelsea McClain, President & CEO, Shaome Cooperative

Hi there, I'm Professor Tola. Now, I may be just a worm, but I've seen things you wouldn't believe. I've been wiggling through soil since the very beginning. Yes, that beginning! Would you like to hear a story about the first Food Forest ever created?

A long, long time ago, I wriggled through the softest, richest soil you could ever imagine. It was called the Garden of Eden, and God created it by himself. There were trees with fruit, nuts, and flowers that danced in the breeze, and underneath the soil, oh the soil; it was alive and happy!

Then one day, I saw something wonderful! God created a man named Adam and a woman named Eve. He gave them a big job: “Be fruitful, multiply, and care for this garden,” God said. He wanted them to grow more food forests, just like Eden! Can you believe it? The world was meant to be full of delicious, beautiful forests!

At first, Adam and Eve listened. They walked through the garden, picking fruit, planting seeds, and caring for the plants. I helped too! I wriggled through the roots, making the soil nice and fluffy for new things to grow.

But after a while, things changed. Adam and Eve disobeyed God. As a result, they had to leave Eden, and outside, they stopped growing food forests. In time, their children and the next generations chopped down trees and left the soil bare. They worked too hard and still grew too little. Oh, how sad the soil and environment had become.

When people forgot their job, the land became dry, tired, and hungry, as did the people.

But I, Professor Tola, kept digging, waiting for someone to remember the food forest. You see, building a food forest isn't magic, it is God's way for provision, naturally! And it all starts with something called **succession.**

“**Succession,**” I whispered, wiggling deeper into the cool, crumbly soil, “is just a fancy word for how nature tells a story—one tiny chapter at a time.”

At first, it doesn't look like much.
Just quiet soil... waiting.

But then, oh! The tiniest green shoots appear. Little grasses and delicate flowers stretch toward the sun, spreading like a soft, living blanket. These are called Pioneer and Perennial plants, and they tuck the soil in, keeping it cool and safe.

Before long, bushes and sturdy Shrubs (or Woody Pioneers) pop up, reaching out as if calling, “Come on in!” Birds flutter down. Bugs wiggle and crawl. Life begins to gather. “The more the merrier,” I chuckled.

Then one day, a gentle shade fell over me. Young trees had begun to grow, their leaves drifting softly to the ground. “Snack time!” I cheered, munching happily and turning those leaves into rich, delicious compost.

As seasons passed, stronger trees stretched their roots deep, preparing the soil for something grand. And then… it happened.

Towering trees rose high, vines curled, fruits and nuts hung heavy, and the forest buzzed with life. I wiggled proudly. “Now this,” I said, “is a food forest.”

A fully established food forest doesn't need to be watered every day or sprayed with strange stuff. It takes care of itself, just like Eden once did!

However, every seventh year the land should experience rest, just as God did on the seventh day. That time of rest should not be overlooked, it allows the soil to heal, restore, and renew itself.

So, kids, if you ever plant a seed, you are doing the work Adam and Eve were first given. You are part of the plan to turn Earth into the fabulous Food Forest it was created to be.

Remember, a food forest isn't a fantasy. It's the way God made the world to work - with trees, plants, birds, bees, worms, and **YOU**!

So, grab a shovel, dig in the dirt, and say hello if you see me. I'll be there, wiggling through the soil, helping the food forest grow.

The End

Or maybe just the beginning...

Bonus Material

Dig Deeper with Professor Tola

Discussion Guide for Classrooms, Homeschools & Curious Minds
Whether you read The Food Forest - Not A Fantasy at school, at home, or under a shady tree, here are some thoughtful questions to help you dig deeper into the story and into God's design for nature!

Think About the Story

1. Who is Professor Tola, and why is he the perfect guide for this story?
2. What made the Garden of Eden such a healthy, thriving place; for people and for soil?
3. What happened when people stopped caring for the land the way God intended?
4. What is a food forest, and how does it reflect God's creativity?

Explore God's Creation

1. Professor Tola teaches us about succession. Can you describe the five stages of how a food forest grows?
2. Why are tiny creatures like worms, bugs, and fungi so important to the soil?
3. What does it mean for land to "rest," and how is that part of God's plan too?

Make It Personal

1. Have you ever helped grow a plant or care for a garden? What did you learn?
2. What are some ways you can help protect and restore the land around you; at school, at home, or in your neighborhood?
3. If you could design your own food forest, what would it include?
4. (Draw it, write about it, or even start building it!)

Bonus Activities

- Draw the Stages of Succession. (Hint: There are five.)
- Write a letter from Professor Tola to future gardeners
- Start a classroom garden or compost bin and watch what happens under your feet!

A Blessing for the Food Forest

Blessed are you יהוה (Yahuah), the God of Abraham, the God of Isaac, and the God of Jacob. You are the King of Heaven and of Earth, who has justified us by faith in Ἰησοῦς (Yahusha / Jesus) the Messiah and commanded us to be fruitful and to multiply the work of your hands. May this food forest become like Eden, like a fruitful vine, with trees planted by streams of living water. May it not wither in times of drought, but drink deeply from the hidden springs you have placed beneath it. May its roots stretch wide into rich, living soil, dwelling in harmony with the plants, the crawling things and the unseen helpers: worms, fungi, and the blessed microbial hosts.

May the seeds planted here bear much fruit; fruit that nourishes, fruit that is sweet to the taste and fruit that lasts.May its fig trees form their unripe figs in their proper season, and the grapevines in bloom give out their perfume in every breeze. May its orchard put forth pomegranates, olives, and other precious nuts and fruits, henna and nard, saffron and aromatic cane, cinnamon and all kinds of frankincense trees, myrrh, aloes, all the best spices. And may you, יהוה (Yahuah), the One who walks in gardens, delight to dwell among this garden.

"To יהוה Yahuah is the earth, and the fulness thereof; the world, and they that dwell therein."
—Psalm 24:1 (The את Cepher)

Caring for the land and environment is part of our purpose and every seed you plant is part of a bigger story.

Glossary of Ecological Terms

<u>Pioneer Plant</u> is a plant that is one of the first to grow in places where no plants have grown before. They make the soil better so other plants can grow later. Examples include moss, grass, and small plants that grow quickly.

<u>Perennial Plant</u> is a plant that grows back every year and produces food again and again, like a friend that keeps coming back. Like Professor Tola, they are the long-term helpers of the forest garden! Examples include asparagus (yummy!), rosemary, and apple or pear trees.

<u>Shrubs</u> (see also Woody Pioneers) are a type of plant that is smaller than a tree and usually have many stems growing from the ground instead of one big trunk. The key traits of shrubs are that they grow low to the ground, have woody stems, often stay green all year, and produce flowers, berries, or leaves that we can use. Examples include Blueberry bushes, Lavender, and Hibiscus.

<u>Succession</u> in a food forest is the natural process of change in the types of plants that grow in an area over time. It starts with simple plants and slowly moves toward a more stable, diverse forest filled with trees, shrubs, and ground plants that all work together. This beautiful process helps create a healthy, self-sustaining system where plants, animals, and people can all benefit.

<u>Woody Pioneers</u> are the first trees or bushes that grow in empty places. They help start the process of building a forest by growing quickly and improving the soil for other plants to come later. These bushes or trees have woody stems or trunks, grow fast and survive in poor soil. They protect the soil and add nutrients. Examples of these include Willow trees, Elderberry, and Alder trees.

About the Author

Lisa Cacao is married to her husband, Javier. She is an educator, homeschooling mother of two, and a 21-year U.S. Air Force retired veteran who serves on the board of Shaome Cooperative, a nonprofit dedicated to food security through education and the planting of food forests and edible gardens. With a background in instructional design, she has created lessons that support social development and student success from elementary through post-secondary levels.

Living in Middle Georgia, at the crossroads of urban and rural life, she connects deeply with homeschooling families, veteran communities, faith-based organizations, and local food-security initiatives. Passionate about regenerative agriculture and biblical stewardship, Lisa writes to equip the next generation with practical tools for faith and action, cultivating a lasting reverence for the Earth and the One who made it.

www.ingramcontent.com/pod-product-compliance
Lightning Source LLC
LaVergne TN
LVHW070626061126
851857LV00002B/24

* 9 7 9 8 9 9 6 3 0 7 5 3 1 *